A Walking Tour of Disney's Zoo

Disney's Animal Kingdom®

A Walking Tour of Disney's Zoo

Disney's Animal Kingdom®

by
Sandi Jerome

SmilingEagle Press Book

Published by SmilingEagle
For information:
SmilingEagle Press
www.smilingeagle.com

ISBN: 978-1-969767029

Printed in the United States of America

Dedication

This book is dedicated to all the animal keepers, doctors, handlers, cast members, and veterinary staff that keep the animals at Disney's Animal Kingdom® healthy and happy and my visits magical.

Thank you to all the Disney cast members and guests for their encouragement including Carley and Crystal Hatley, Bex Britz, Maribel Hermida, Kayla and Kathy Montizaan and Keirsten Sangster.

A Walking Tour of Disney's Zoo

Disney's Animal Kingdom®

Prologue

I grew up on an avocado and orange ranch in San Diego County where my family also raised chickens, rabbits, cattle and pigs. I raised sheep as a 4-H project and had dozens of cats, dogs, and even peacocks as pets. I wanted to be a large animal vet when I grew up. I had plans to attend California Polytechnic State University,but when my lamb got sick and I had to give it a shot, I fainted. I became a technical writer, programmer, and software developer instead.

My grandparents lived a mile away from Disneyland which was being built the same year I was born. Growing up, when my family would go to Disneyland, they couldn't tell me about their plans because I would throw up in excitement. Again, my weak stomach. Instead, my parents would keep it a secret and at the last minute get off one exit earlier than my grandparents' house. Each time we visited my grandparents, I would watch and watch to spot the Disneyland Matterhorn in the distance, then wait to see if we were going to the Happiest Place on Earth. Since my parents couldn't afford to visit Disneyland often, it was unusual for them to make that exit. But each night that we visited my grandparents, I would struggle to stay up late enough to see the fireworks from their front yard. Animals and Disney; it was almost like they built Disney's Animal Kingdom for that little girl who loved all things Disney and animals.

A few years ago, we moved to the Treasure Coast area of Florida to be close to our four granddaughters and help raise them. We truly enjoyed sharing our love of Disney as they grew up. After they were grown, we retired and moved closer to Disney World and became annual passholders. Today I can visit all the parks as often as I want and Disney's Animal Kingdom is the one closest to our home. We were one of the first subscribers to Disney+ and I was delighted with the new show, National Geographic's "Magic of Disney's Animal Kingdom." After each episode, I wanted to know more about these amazing animals and started my research.This book is an homage to all the animal keepers, doctors, handlers, cast members, and veterinary staff that keep the animals at Disney's Animal Kingdom healthy and happy and my visits magical. We have recently watched the filming of the 2nd season of the show due to air in late 2022.

I try to spend as much time as I can at Disney's Animal Kingdom. With my weak stomach, I don't ride many thrill rides and instead have found a delightful way to enjoy this park without staring at my phone to check wait times. If I can't go to the parks, I stay updated by watching my favorite YouTube vloggers listed in my Dedication.Their videos let me get ride all the most exciting rides with them and I never get sick.

Introduction to Disney's Animal Kingdom

Over 200 species of animals are living on the vast 540 acres of Disney's Animal Kingdom. Some might think of it as a theme park with thrilling rides, but to me, it is a world-class zoo. Disney uses things like moats, water features and carefully disguised fences (usually in the water-filled moats) to keep the animals in their habitats - and the humans out. These "fences" make it appear that the animals are roaming free in their natural environment. Often these habitats need refurbishment, so the animals are moved to new locations. In addition, at night and during the dangerous Florida storms, they move them into their barn shelters "backstage." For some animals, like the anteaters, Annie and Callie; there are "doubles" of the animals. One anteater comes out in the morning and goes back into her enclosure around noon; switching places with her double. Animals also need medical checkups, so your favorite animal might be missing today for a variety of reasons. Disney frequently moves the animals around or removes them from the exhibit completely. During Covid, they moved porcupines away from Discovery Island because the drums from the cavalcade boats were too loud. We hope they come back soon.

This book is mostly about the mammals in the park with a few reptiles too. I am working next on *Birds andFish of Disney World,* and *Plants, Trees and Flowers of Disney World.* I have sketched most of the animals to make spotting them easier and encourage you to do the same. I'm not an artist by any stretch of the imagination, so I've set the bar pretty low for your own sketching. You might want to take a sketch pad with you. At the end of each description, I included the scientific name. Animals are given scientific names because it allows people around the world with different languages to communicate accurately about animal species.

My favorite hot Florida day is getting up early and following this walking tour of the beautiful shady park with a few breaks for delightful treats. At the end of this book, I mention my favorite places to eat; but while I'm walking, my favorite treats are the Dole Whip atTamu Tamu Refreshments (Africa) and the Tiger Tail Chocolate Twist at Isle of Java (Discovery Island.) The Isle of Java is also where you can get the breakfast biscuit without sausage. I like to take my breakfast biscuit next door to the Flame Tree Barbeque seating area and sit down close to the lake where it is quiet and cool in the early morning before that quick service restaurant opens.

Are you ready to get walking? I hope so. This is a fun tour with some of my amazing animal friends at Disney's Animal Kingdom.

Walking Tour of the Animals

This guide is meant to be a walking tour of the animals and concludes with the Kilimanjaro Safari and then Dinoland. It will take you through the Oasis starting with the Reeve's Muntjac Deer, then Discovery Island, Asia, Africa, and the safari ride. At Dinoland you'll see the American crocodile and that is where you can meet the others in your party who might be on the thrill rides.There are lots of places to sit including Restaurantosaurus that is warm in winter and cool in summer.

If you are trying to fit this into a full Animal Kingdom visit and using Genie+ and Lightning Lanes; you might schedule those premium rides around your walk. This gives you something to do while waiting for your next Lightning Lane (former Fast Pass.) But if your objective is to merely see the Animals, then you only need to schedule the Kilimanjaro Safari towards the end of the visit - or first thing, depending on the Lightning Lane schedule. As a bonus feature, when the park gets very busy, you can take a rest by taking the bus to Disney's Animal Kingdom Lodge and tour their savannah and have a lovely meal at one of the restaurants at the lodge. The quick service, Mara is fantastic and you can take your flatbread sandwich "to go" and eat it by the pool/spa area and watch the giraffes walk by.

The full park is almost 8 miles of walking, but there are no animals in Pandora or most of Dinoland. This means your walk will be less; about 3-4 miles. Even though it is one of the biggest parks, many of your views of animals are from the Kilimanjaro Safari ride. Taking the safari ride lets you rest, and even though you might have to spend time in line, you can read this guide while waiting. Another available rest is when you take the Wildlife Express Train at the Harambe Station toRafiki's Planet Watch. The Affection Section has a "hands on" experience with some animals. If they are offering it, spend the time doing The Animation Experience at Conservation Station. That is a nice place to get out of the rain, winter cold, or summer heat. If you are not interested in drawing, then hang out at the hospital window and you might see some of the Disney+ stars like Dr. Dan or Dr. Natalie giving the animals a checkup or performing surgery. One day while I was taking my walking tour, the meerkats were missing. They are some of my favorite animals to watch. The keeper explained that they were up at Rafiki's Planet Watch. Since they are very social animals, when they bring one up for a checkup – they bring them all to prevent future exclusion by the group. I was able to watch one of meerkats get an exam. So, if your animal is missing; ask the keepers and/or cast members and you'll probably get an explanation. Other animals are shy and smaller than you would think. In the Oasis, notice that the Reeve's Muntjac Deer and Wallaby are small; only a few feet tall. You really need to look hard to spot them.

The Oasis - Right Trail

If you think "right" - that might be helpful to get started. The security line on the right is usually shorter in the morning because guest parking and resort buses are on the left. Before you enter the park there are restrooms on your right and a cash machine. If you are a passholder, that special entrance is usually on the right. After you enter the park, head to the right because most people go to the left to visit Pandora first. If you need an ECV or stroller, those rentals are on your right along with a small gift store. This "after ticketing" area is called the Oasis, so head to the trail on your right. After you start on that trail, your first animal is on your left, slightly uphill from the stroller rental pickup. Look for the sign for the Reeve's Muntjac Deer.

Map of the Oasis

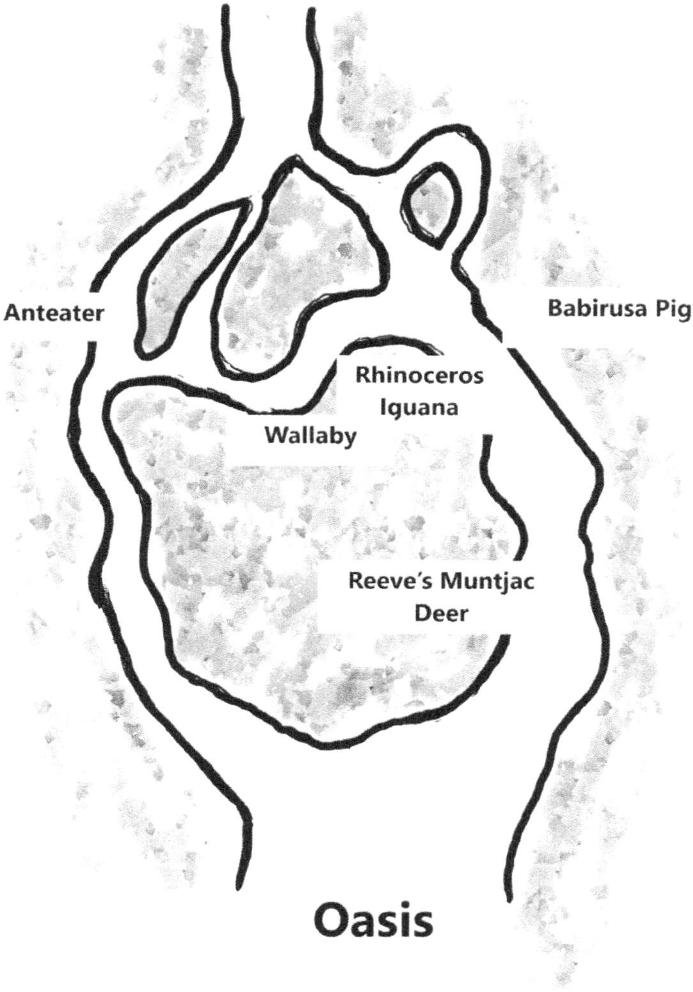

Reeve's Muntjac Deer

Bambi lovers will love these cute, tiny deer. They can also be one of the hardest animals to spot in their viewing area; they are a lot smaller than you would think and they often are hidden in the bushes. Going up the right side of the Oasis, right before the babirusa pig, they are on the left side of the trail. Look for the sign, then look for a very small animal, often hidden in the bushes.

These deer are only about one foot high at the shoulder, and three feet long. They have striped markings on their face. Their belly is lighter; almost white. That lighter color fur extends up to the chin and is also on the bottom side of the tail.

Only the males have antlers; about four inches long. The males will use their antlers to push enemies off balance and then bite them with their two-inch upper canine teeth. The Reeve's Muntjac got its unusual name from an employee of the British East India Company, John Reeves. It can be found mostly in southeastern China and in Taiwan although it has now been introduced to some European countries and Japan.

The Reeves' muntjac is sometimes called the barking deer for the sound it makes when provoked. It likes to eat grasses, berries, blossoms, herbs, shoots, nuts and sometimes eggs or tree bark. It prefers a solitary life and is active at dawn or dusk. The scientific name is for the Reeves' Muntjac Deer isMuntiacusreevesi.

Babirusa Pig

If you cross over the trail to the other side, you'll see the sign for the Babirusa Pig. With its distinctive snout and curved back, it is easy to see that the babirusa is part of the pig family. The name, babirusa, means "pig deer" in the Malay language. They got this name because of the long graceful deer-like legs and the males have curving tusks that look like deer's antlers. These tusks can grow up to 17 inches long.

Disney's Betty the Babirusa became a star in episode 3 of Disney+'s "Magic of Disney's Animal Kingdom," from National Geographic when they bred her to Mentari from the Tampa Zoo. About 155 days later, a darling new babirusa piglet, Kirana, was born. Betty and Mentari were chosen to breed through the Species Survival Plan (SSP) overseen by the Association of Zoos and

Aquariums. The SSP assures responsible breeding for endangered species, such as the babirusa pigs. The exciting birth of Kirana which translates into “sunbeam” in Indonesian was quite an accomplishment for Disney’s Animal Kingdom.

Related to the farmer’s domestic pig, babirusas differ because they have complex two-chambered stomachs like sheep. They are omnivorous and will eat almost anything; berries, leaves, nuts, brush, bark and even small animals. Babirusas are native to Sulawesi, an Indonesian island east of Borneo and like to live in the rainforest and swamps. Unlike pigs that you see at the county fair that have ten or more piglets, the babirusa only has one to three piglets. Some think that the babirusa might have traveled to other Indonesian islands by swimming, but man could have brought them since they were bestowed by native royalty as a gift. They are endangered today and in Indonesia killing them is illegal.

Betty and her daughter, Kirana are hard to spot because there is quite a bit of foliage in the enclosure. You can see Mentari on *Secrets of the Zoo: Tampa - Creature Feature: Sulawesi Babirusa.* He got his long tusks trimmed for his upcoming date with Betty.The scientific name for the Babirusa Pig is Babyrousacelebensis, babyrussa.

Oasis Trail Crossing - Left to Right

Instead of continuing up the hill after the babirusa pig, cross over to the left again and you'll see an opening for another trail that goes into the center of the Oasis. There are many birds on this trail, but those will be covered in another book.

Discovery Island Trails - Left to Right

After you pass the Cotton-top Tamarin Monkey on your right, you'll be back on the left Discovery Trail that crosses the island again. There are two trail entrances; one right at the monkey and another right before the bridge into Africa. Take either one and you'll be heading back into Discovery Island and towards a beautiful waterfall at the exit to "It's Tough to be a Bug." On your right are two exhibits; the Galapagos Tortoise and Porcupine.

Porcupine

At the time of this book, the porcupine that was on Discovery Trail had been moved back inside its barn because the drums on the cavalcade boats had disturbed it. It is listed as an African crested Porcupine.

But I got to see Peri, a prehensile-tailed porcupine and her new baby girl, Shelley, on the "One Day at Disney" documentary on Disney+. The Prehensile-tailed Porcupine is native toCentral and South America. She was also featured in episode 6 of Disney+'s "Magic of Disney's Animal Kingdom," from National Geographic where she gave birth to a darling baby girl. The baby was named after Veterinary Operations Manager, Shelley.

The African crested porcupine is native to Italy, North Africa and sub-Saharan Africa. It is 2-3 feet long and weighs between 30-60 pounds. Almost the entire body is covered with coarse bristles that are dark. Along the head, nape, and back are sharp quills that can be raised into a crest, hence the name crested porcupine. There are even sturdier quills over a foot in length that run along the sides and back half of the body. These sturdier quills are used for defense and have light and dark bands. These quills are not firmly attached. This porcupine has a shorter tail which has rattle quills at the end.

In the spring and summer, porcupines eat berries, seeds, grasses, leaves, roots and stems. In the winter they primarily eat evergreen needles and the inner bark of trees. The scientific name for the African Crested

Porcupine is Hystrix cristata and for the Prehensile-tailed Porcupine is Coendou prehensilis.

Galapagos Tortoise

The Galapagos Tortoise is the largest living species of tortoise and can be over 900 pounds. As one of the longest living vertebrates, they can live over 100 years and one captive Galapagos tortoise lived 175 years. This slow-moving animal can best be seen on the Discovery Island trail if you turn onto the trail across from Creature Comforts. The tortoises are herbivores and eat leaves, grasses, fruit like berries, melons, and oranges. Young tortoises can eat over 15% of their own body weight with a digestive system of hindgut-fermentation similar to horses.

Tortoises get most of their water from the dew and sap in vegetation so they can survive longer than 6 months without water. They can survive for up to a year without food and water by breaking down their body fat to produce water as a byproduct. Tortoises have slow metabolisms but when thirsty, they may drink large quantities of water very quickly, storing it in their bladders and neck.

The tortoises have a shell of brown or grey color that can be various shapes from domed to saddleback. The plates of the shell are fused with the ribs in a rigid protective structure that is integral to the skeleton. Each tortoise has a pattern on their shells throughout life and can withdraw its head, neck, and fore limbs into its shell for protection. The legs are large and stumpy, with dry, scaly skin and hard scales. The front legs have five claws, the back legs four.The tortoises are cold-blooded, so they bask for a few hours in the morning to absorb the sun's heat through their dark shells before eating for most of the day. They are native to the 7 Galapagos Islands that are part of the Republic of Ecuador - 563 miles west of Ecuador. The scientific name for the Galapagos Tortoise is Chelonoidis nigra.

As you continue on the trail, you'll go past the exit for It's Tough to be a Bug and then be back on the main walkway. On your left is another animal added in 2022.

Black Mountain Tortoise

This tortoise can be hard to spot, but I finally found the Black Mountain Tortoise hiding under a low branch of a tree. Native to the forest of northwestern Thailand to northeastern India, it is thought to be among the most primitive of living tortoises. Large adults can weight up to 55 pounds. They eat mostly grasses and other plants, but will eat insects and frogs. The front legs are very large and pointed with dragon-like scales. The scientific name for the Black Mountain Tortoise is Manouriaemysphayrei.

After this exhibit, turn left and cross over the bridge into Asia.

Rhinoceros Iguana

You'll find this beauty often sunning on a rock on your left as your start on this interior Oasis trail. Look for the sign for the Rhinoceros Iguana.The Rhinoceros

Iguana is a large lizard is native to the Caribbean Island of Hispaniola (Haiti and the Dominican Republic.) They can be close to 2 to 5 feet in length and their skin is anywhere from a grey to dark green and sometimes brown.

They like to eat a variety of plants, flowers, and fruits. But they will also eat insects, lizards and even snakes if the opportunity arises.The scientific name for the Rhinoceros Iguana is Cycluracornuta.

Wallaby

A few feet further on the trail, on your left is the wallaby, which might be hard to spot. My first sighting was after a cast member pointed the little wallaby out to me. They are only about 18 inches tall and weigh 15-20 pounds. This tammar wallaby is the smallest of the wallaby species. Native to South and Western Australia, the tammar wallaby is grey in color. They have a small head and large ears. Their long tail is thick at the base. They have the ability to retain energy while hopping; each hop returns energy back into its tendons when it lands in a recoil effect.

In 2014 a new baby joey was born and another in 2016 at Disney's Animal Kingdom. The tammar wallaby is most active at night, eating mostly grasses. They also eat plants; up to 24 different plant species and they can drink seawater.The scientific name for the Tammar Wallaby is Macropus eugenii or Notamacropuseugenii.

By now, you might think you've seen all the animals in the Oasis, but there is one more; the anteater. Exit the trail and you'll be on the main left Oasis trail. The anteater is on your left.

Southern Giant Anteater

Most of my views of the giant anteaters at Disney's Animal Kingdom have been of Annie's or Callie's bushy tail as she is curled up in the far-right corner sleeping. There are two of them and you can tell them apart when they are walking around because Callie has white boots. You most likely won't see them together; only one comes out on display at a time. The anteater likes to sleep during the heat of the day with her huge tail covering most of her body. This native of South America lives in the Amazon basin and the grasslands of Paraguay and Argentina, but can be found as far north as Mexico. With that long nose, the anteater can smell ants and as the sign says by their enclosure says, "can gobble up as many as 30,000 bugs a day!" Their tongue is extremely long; greater than the length of their head. They don't have any teeth but use those claws to dig into termite mounds. That lovely fur protects them from the counterattacks by insects. With humans around, the anteater tends to be more nocturnal and can live to be 25 years old. They have one of the lowest body temperatures of any mammal. They can turn the heat down to stay cool during the hot days, but turn it up at night when they go hunting for bugs. The scientific

name of the Southern Giant Anteater is Myrmecophaga tridactyla.

After seeing the anteater, your next destination is to go towards Discovery Island. You have two options; go back into the interior Oasis trail where you came from and you'll notice a bridge on your left that leads into a big rock. It is a fun secret trail that will take you into the main walkway into Discovery Island. The 2nd option is to merely follow the flow of people after you see the anteater up the left main Oasis trail and onto the bridge to Discovery Island. When you reach that bridge, there is often a long line to get that amazing "Tree of Life picture." If the line is short, then get that keepsake picture now.

Wilderness Explorers

You might want to start your tour by getting a Wilderness Explorer's Guide that contains self-guided activities ranging from animal observation to learning important wilderness skills. There are over 25 badges to earn and almost a dozen locations throughout the park to learn more about the animals. The first booth is on the right side of the bridge from the Oasis to Discovery Island, but might have a long line or it is too early for someone to be there. You can get the guide at any of the other locations throughout the park and it is a nice guide and a chance to interact with the cast members at these tables.

Tree of Life and It's Tough to be a Bug

Directly in front of you is the Animal Kingdom's landmark item, the Tree of Life. This 145-foot-tall man-made baobab tree has 325 carvings and it took 10 artists and 3 Imagineers working full time over 18 months to create. We'll be sort of winding around it on the Discovery Island trails; getting a true up-close look at the carvings. Our walking tour continues to the left of the stroller parking for It's Tough to be a Bug.

Discovery Island Trails - Right to Left

These cool and shaded trails inside Discovery Island are one of the best-kept secrets in all of Disney's Animal Kingdom. It is easy to miss the entrances to the trails. Look for the two signs about the lemurs.

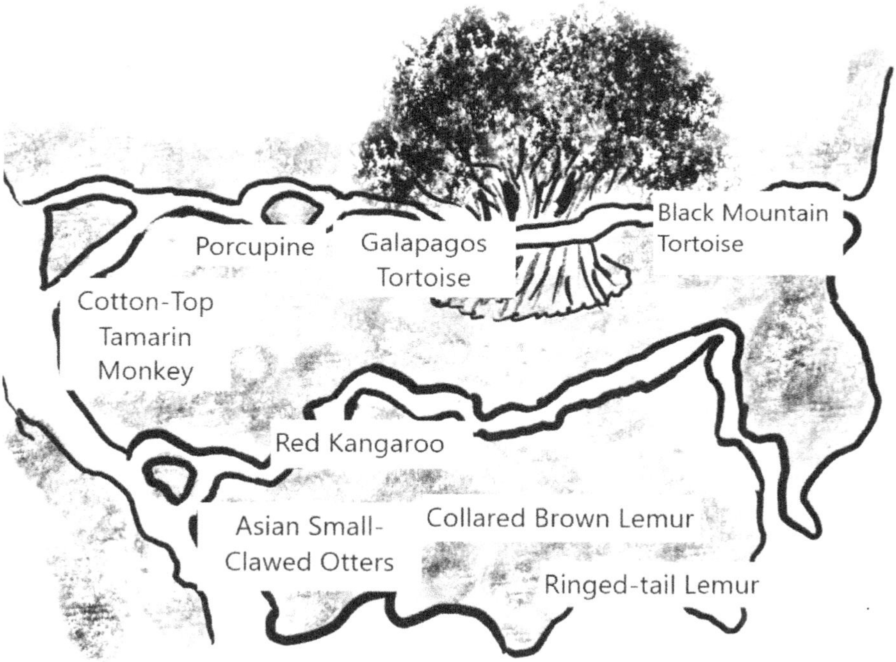

Discovery Island

Map of Discovery Island

Collared Brown Lemur

Both the collared Brown and ringed-tail Lemurs are in the same area of Discovery Island, just to the left of the stroller parking for It's a Bug's Life. The brown lemurs are called the true lemurs and have reddish brown fur compared to the ringed-tail. but the ringed-tail is easier to spot with its bright black and white ringed tail.

The male Collared brown lemur can be distinguished by their cream-colored or reddish-brown beards and darker head. The Collared brown lemur is usually found in multi-male/multi-female groups of two to over a dozen lemurs. Female dominance, which is common in other lemurs, is not found as much in the collared brown. The scientific name for the Collared Brown Lemur is Eulemurcollaris. There is another viewing area as you head up the Discovery Island inner trail, staying to the left of the line into It's a Bug's Life.

Ringed-tail Lemur

The Ring-tailed lemur lives in larger groups of closer to thirty with female dominance. They are vocal, often alerting others to a predator. The Ring-tailed is a larger lemur, weighing close to five pounds. Their tail is longer than their body and is not prehensile. Instead of being able to grasp things, it is merely used for balance and communication. They have dark racoon-like circles around their eyes, a dark muzzle and a black nose. They have slender fingers with human-like nails. Their thumb is shorter and separated from the other fingers and opposable to enable them to rip open fruit and seeds.

The lemurs eat fruit, seeds and leaves, from a wide variety of plants. The collared brown lemurs eat from almost a hundred different plant species. The ring-tailed are known to eat bugs during the dry season when fruit is not readily available. Similar to meerkats, they will sit in a lotus, sun-facing position in the morning to get warm.

Lemurs are native to Madagascar, an island off the coast of Africa. The scientific name for the Ringed-tail Lemur is Lemur catta.

There are three viewing areas for the lemurs on your left. Continue along the trail as you walk next to the roped off line for It's Tough to be a Bug. This is a wonderful close-up view of the Tree of Life. You'll pass under a waterfall rock and after a couple curves, start looking for the kangaroos on your left. There isn't a sign here; the kangaroo sign is in the main Discovery Island area by the otters and flamingos - but it is a great sighting area. You might need to stand on your tippy toes to see them over the rocks before the trail bends to the right and puts you back on the main left trail.

Red Kangaroo

It might be hard to spot a "red" kangaroo since only the male is red. Most of my sightings have been of the smaller female kangaroos that are browner or grey in color. The males are much larger and can grow to 200 lbs. and over six feet tall. They have been known to jump over 25 feet with their strong hind legs functioning like a rubber band to propel them.

They are native to most of Australia and one of the largest mammals and kangaroos in Australia. The red kangaroo eats green grass and the succulent part of herbaceous vegetation, feeding mainly at night. They can live for long periods without water, utilizing the moisture in their food. Although the sign for the red kangaroo is next to the otter exhibit at the left side of

Discovery Island (your next stop) these are best seen from inside the Discovery Island Trail. The scientific name for the Red Kangaroois Osphranter rufus.

You have to backtrack a little to see the last Discovery Island animal. As you head toward the main trail, you'll see a bridge and a catfish, but birds and fish will be covered in a future book. Turn left, back towards the entrance and the otter is on your left.If you didn't see the kangaroos when you crossed Discovery Island, their sign is on the left by the flamingos.

Asian Small-Clawed Otters

The otter complex has a couple viewing areas on Discovery Island. As you go down the steps in the area closest to Creature Comforts; you'll be able to see them both under water and on one of their islands in the exhibit. The next entrance shows them on their other island.

These playful otters are native to Indonesia, Philippines,and southern India and China. In the 1980s, a few Asian small-clawed otters escaped from captivity and established a population in the wild of England. They are called small-clawed because their claws don't extend past the pads on their webbed feet. When swimming on the surface, they row with their front paws and paddle with the rear paws. When diving underwater, they swim with an up-and-down motion with their bodies and tails.

They like to eat mollusks, crabs and other small aquatic animals. In the hot summers, the Disney keepers feed them fish popsicles; fish frozen in a thick layer of ice as a fun and cool treat.

They typically live in pairs and sometimes in a small group of about a dozen in riverine habitats, freshwater wetlands and mangrove swamps. The Disney keepers have taught them to open their mouths for easy dental checkups. They have darker brown fur that is lighter below. Their cheeks and neck can be almost white. They have small oval-shaped ears. They are the smallest otters in the world, weighing about ten pounds. The scientific name of the Asian Small-Clawed Otteris Aonyx cinereus.

After you come out of the Otter exhibit, you're back to the "main drag." Head away from the entrance and instead towards Africa. The Creature Comforts - the

Starbucks inside of Disney's Animal Kingdom should be on your left. On your right, you'll see an exhibit for the cotton-Top tamarin monkey.

Cotton-Top Tamarin Monkey

The cotton-top tamarin monkey is native to the tropical forest forests in northwestern Colombia. In the area across from Creature Comforts on Discovery Island, you can find them in their exhibit; often way up high. They are smaller than you would think; only 15 ounces and 15-18 inches high. Their dark faces stand out against white top fur. They have fur covering most of the body except the palms of the hands and feet. Their back is brown, and the underparts, arms, and legs are whitish-yellow. They have sharp nails that help them climb trees.

The cotton-top tamarin prefers to eat fruit, insects, and fluids from plants such as gum, sap and nectar. It will sometimes eat reptiles and amphibians. Due to its small body size and quick food digestion, this monkey's food must be high-quality and high-energy. The scientific name for the Cotton-top Tamarin Monkey is Saguinusoedipus.

Asia Walkway

A great portion of Asia is taken up by the main attraction, Everest, but Asia also has two amazing animal areas; the Gibbons exhibit and the Maharaja Trail.

Gibbon

There are two species of gibbons in the area close to the entrance to the Kali River Rapids and Maharajah Jungle Trek. The darker ones are the siamangs and lighter gibbons are the white-cheeked gibbons. You will often hear them before you see them as they howl to each other. A large throat pouch can be inflated to the size of the siamang's head, allowing it to make loud, resonating calls or songs.

The siamang is native to the forests of Indonesia, Malaysia, and Thailand. This gibbon is mostly black and large; weighing 30 pounds and 3 feet tall. The siamang eats a large variety of plants, flowers and insects. Its major food source is figs. The siamang prefers to eat ripe figs and younger vines and leaves. The white-cheeked gibbons are native to Laos, Vietnam and southern China. They live in evergreen tropical rainforests and monsoon forests. Like the siamang, they prefer ripe fruit or insects. But they are smaller than the siamang, only reaching about 2 feet and weighing 15 to 20 pounds.As babies, the white-cheeked gibbons are tan in color but gradually turn darker as they age. The scientific name of the Siamang Gibbon is Symphalangussyndactylus and the White-cheeked Gibbon is Nomascusleucogenys.

Maharajah Jungle Trek

This trail in Asia is right past the entrance to the Kali River Rapids. It is about a third of a mile and should take 15-20 minutes to enjoy. The highlights are the tigers and Komodo dragons.This trek was designed to look like ancient Indian ruins.Disney's famous imagineer, Joe Rhodeused as an inspiration for these ruins the architecture and history of Nepal, Thailand, Indonesia and India. Notice the signs, photographs, colors, buildings, animal carvings. as they appear throughout the trail. This is a place in Disney's Animal Kingdom that lets you truly experience being somewhere else in the world due to the amount of theming along the trail. Animals do come and go and the area that houses the Lion-Tailed Macaques used to host Malayan Tapirs until 2010.

Map of Maharajah Jungle Trek

Komodo dragon

There are three Komodo dragons at Disney's Animal Kingdom. A female, Tiamat or Tia, a larger male named Deidara and a smaller female,Khalessi. The two larger dragons are switched out of the exhibit on the Maharajah Jungle Trek and the little one is usually up at Rafiki's Planet Watch across from the petting area.

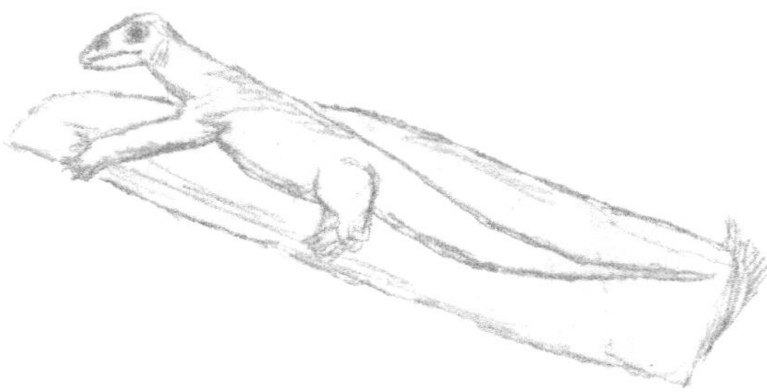

The Komodo dragon is native to Indonesia and can be found mainly on Flores, Gili Motang, Rinca and of course Komodo Island. As the largest of the lizards, they are normally the apex predator in their habitat and eat deer and decaying animals. They will even attack humans and have venomous bite. You can see them safely on the left of the Maharajah Jungle Trek or by venturing up to Rafiki's Planet Watch on the train.

Komodo dragons usually weigh around 150 pounds and have tails as long as their body. They have about sixty sharp and serrated teeth that can be one inch long. Their skin has scales, which contain tiny bones that function as a sort of natural armor. The only areas lacking these protective scales are around the eyes, nose and mouth. There is a light-sensing organ on the top of its head. They have a long, yellow, deeply forked tongue.The scientific name of the Komodo dragon is Varanus komodoensis.

Lion-Tailed Macaques

The lion-tailed macaques are of the old-world monkey family and are named lion tailed because of the big white and gray mane surrounding their face and their lion-like tail with a tuft at the end. The tuft is larger on males than females. Their face and fur are both black and the face is hairless. The lion- tailed macaques weigh around 20 pounds and are among the smallest of all macaques.

These three sisters at Disney's Animal Kingdome are from Germany. They are named Dorothy and Emily, and another is a German name - which means One without a Tail. You might be able to see why. They smack their lips to communicate with each other along with many different vocalizations including loud cries when threatened.

They like to eat leaves, fruits, tree buds, insects, small vertebrates, seeds, shoots, flowers, cones, and have been known to prey on the eggs of pigeons. They are native to the evergreen forests of Western Ghats in South India. You can find them on the Maharajah Jungle Trek, right past the Komodo Dragon on the left. There are two viewing areas before you enter the bat house. The scientific name of the Lion-Tailed Macaquesis Macaca silenus.

Bats

Although they look like birds, bats are actually mammals. The bats can be found on the Maharajah Jungle Trek in a special "house" that you enter after viewing the macaques. They have two types; the Rodrigues fruit bat which is native to the Indian Ocean Island of Rodrigues in Mauritius and near Madagascar. The other is the Malayan flying fox; native to southeast Asia.

The Malayan flying fox is one of the largest species of bats and also known as the large fruit bat. It actually doesn't use echoes to "see" - but instead has good eyesight. It feeds on fruit, flowers, and nectar, but prefers flowers and nectar. When they do eat fruit, the flying fox prefers the pulp, and slices open the rind to get it. It weighs about 1-3 pounds and has a foxlike face with pointed ears.

The Rodrigues fruit bat is much smaller, weighing less than a pound. It eats the fruit of various trees, such as figs, mangoes, and palms. Like Malayan flying fox, they squeeze out the juices and soft pulp, rarely swallowing the harder parts. The scientific name for the Malayan Flying Fox is Pteropusvampyrus and for the Rodrigues Fruit Bat, Pteropusrodricensis.

Sumatran Tigers

The Sumatran tiger is one of the smallest tigers and tends to have more stripes than the other tigers.The Sumatran tigers are from the Indonesian island of Sumatra and are the only surviving tiger population in the Sunda Islands of Indonesia, where the Bali and Javan tigers are already extinct. It is also called the Sunda Tiger. They are considered to be Critically Endangered with less than 1,000 left in the world.

These tigers at Disney's Animal Kingdom can be hard to spot, they prefer forests with dense understory cover and steep slopes and they strongly avoid forest areas with humans.They like to use areas closer to water and prefer areas with older plants, more leaf litter, and thicker subcanopy cover. Tigers mainly eat deer, wild pigs, water buffalo and antelope. I am sure that the water buffalo that you might spot behind their enclosure is very tempting to these tigers.

Back in 2017, Disney welcomed two new tiger cubs at Disney's Animal Kingdom, twins Anala and Jeda. Now, Jeda has been relocated to San Antonio Zoo as part of the Species Survival Plan (SSP). The parents were Sohni and her visiting mate Malosi. Anala and Jeda were the first Sumatran tiger cubs born at the park and when I visited, Sohni and Anala were able to be seen in different places on the Maharajah Jungle Trek. You might want to ask a cast member for help in spotting them in the multiple viewing areas along the trail. In episode 4, of Disney+'s "Magic of Disney's Animal Kingdom," from National Geographic, Sohni undergoes a voluntary blood draw. In episode 6, daughter Anala gets some enrichment to keep her hunting skills sharp. We are hoping that next season, we'll see the new male Conrad that arrived in 2022 after Anala was relocated as part of the SSP.

In 2025, a new cub was born named Bakso. There were long lines to see this darling new tiger.

The scientific name of the Sumatran Tiger is Panthera tigrissondaica.

Asian Water Buffalo

There are three main types of water buffalo; the domesticated, African or Asian. The ones in Africa can be dangerous because they are rather aggressive. These lovely girls are the gentler Asian Water Buffalo and are considered endangered. The bigger ones' names are Blanche, Dorothy, and Rose. They came in 2015 to Disney's Animal kingdom where a new pond was added since the water buffalo have few sweat glands and they like the water.

They are called water buffaloes because during floods, they eat underwater in rivers and swamps. In the wild, the water buffalo eats reeds and grasses. The ones you see on the Maharajah Jungle Trek are probably eating hay or alfalfa. The smaller water buffalos in the exhibit are names Sophia, Betty Bea and Rue. You might remember all these names from the popular TV show, Golden Girls.

Although originally from Southeast Asia, China and India, they can now be found in North America, South America, Africa, Australia and even in Europe. Most of the world's population of water buffalo are in Asia. They were introduced to the United States in 1974 to study at the University of Florida. These black-skinned bovines typically weigh about 600 to 1200 pounds, but some have been found to get closer to 2400 pounds. Their horns grow downward and backward, then curve upward in a spiral. They have a long neck and their tail reaches down to their hocks.

You can first spot the water buffalo on the Maharajah Jungle Trek, on the left side, behind the tigers. You'll see them again as you come around to a circular clearing. The scientific name of the Asian Water Buffalo is Bubalus arnee

The Maharajah Trek ends in an aviary. If you want to "bypass" the birds, there is a sign with an arrow to the left to go around and skip the birds. I will cover the birds in a future book. At this time, you might be getting tired after walking from your car to the entrance and then doing half of the Oasis and Discovery Island Trail. After exiting The Maharajah Trek you're now at one of my favorite Disney quick service Restaurants on your right; Yak and Yeti Cafe. Your next stop is Rafiki's Planet Watch that doesn't have any food, so the Yak and Yeti Cafe or the Harambe Market before the train is a great place to grab some food.

Rafiki's Planet Watch

On the Wildlife ExpressTrain up to Rafiki's Planet Watch, you'll pass the various "barns" for some of the big animals like the elephants and rhinoceros. After arriving, there is a long winding trail with a variety of activities with special interest for the Wilderness Explorers. At the end of the trail are some restrooms on your right and a huge animal mural. Look for some hidden mickeys in a butterfly.

Animation Experience

As I mentioned earlier, I'm not an artist, but I did the rough sketches in this book to help you spot the animals. If you are interested in sketching, be sure to take a lesson at the Animation Station where you can sketch one of the many animated Disney animals. Each 25-minute class usually offers something different. They can fill up quickly, so if this is something you want to do, then make sure you check those class times first.

Conservation Station

The Conservation Station includes numerous exhibits; a nutrition center where you can watch animal meals being prepared, various displays of reptiles, insects, snakes and a viewing window into a veterinary treatment room. This building is a great place to spend your time during thunderstorms. Most of the animals are in the Affection Section which is a hands-on petting zoo, except the aardvark and Komodo dragon.

Aardvark - African Ant Bear

I first met Willie, Disney's ambassador aardvark, at Rafiki's Planet Watch. His handler had him on a leash and was walking him around. He became famous a few years later in episode 5 of Disney+'s "Magic of Disney's Animal Kingdom," from National Geographic called "Aardvark Love" when he was matched up with a female aardvark, Peanut. About seven months later, the very cute baby aardvark, Karanga was born. Since she was the first birth of an aardvark in the park's history, the Disney keepers decided to name her after her mom, Peanut. Karanga means "Peanut" in Swahili. In the wild, the young will stay with their mother for about six months.

Aardvarks are native to Sub-Saharan Africa and like to root and burrow in the dirt. Similar to the anteaters, they can eat 50,000 insects in one night. Willie isn't picky, he'll also eat bananas and avocados. On occasion, his Disney handlers will give him honeycomb larvae as a special treat. Aardvarks live a little over twenty years. Although their name in Afrikaans means "earth pig" or "ground pig" they are not closely related to the pig; rather they are more related to the elephant. Looking at their nose, I can understand why. They swing their long noses from side to side to pick up the scent of their favorite food; termites. They use their long and sticky tongues to scoop up the insects. Their ears are also long and tall like a kangaroo, but they are not related to them. In Africa, they like to use their claws to burrow into the earth during the day to stay cool. The scientific name for the Aardvark is Orycteropus afer.

Komodo dragon Viewing Window

There are three Komodo dragons at Disney's Animal Kingdom. A female, Tiamat or Tia, a larger male named Deidara and a smaller female Khalessi. The two larger dragons are switched out of the exhibit on The Maharajah Jungle Trek and the little one is usually up at Rafiki's Planet Watch across from the petting area in a viewing window. There is more info on the Komodo dragon under the Maharajah Jungle Trek section.

Affection Section - Petting Zoo

This delightful area lets you get upfront and personal with some cuddly animals. You'll find sheep, goats, and even chickens. Episode 5 of Disney+'s "Magic of Disney's Animal Kingdom," from National Geographic, featuredPopcorn, a rescued chicken. Popcorn required surgery to remove an impacted egg. I won't go into detail about these animals because they are often rotated out, but the one that is interesting is the KuneKune Pig, native to New Zealand. It has remarkable social learning skills with an astonishingly good memory.

Gorilla Falls Trail

When you get out of the train area, turn to the right and you'll immediately be going up the hill to the entrance of the Gorilla Falls Trail. If the wait time for the Kilimanjaro Safari isn't long, you might want to ride that first because it also ends into the entrance of the trail, but I find thatI'm rested from the train ride (and waiting for the train) and ready to keep walking.

Map of Gorilla Falls Trail

Angolan Black and White Colobus Monkey

On the right after you start on the Gorilla Falls Exploration Trail, there is the habitat of the very unusual looking Angolan black and white colobus monkey. During episode 7 of Disney+'s "Magic of Disney's Animal Kingdom," from National Geographic, we got to meet Alika who was soon to become a big sister to new baby Douglas born to the same mother, Zahra.

Angolan black and white colobus monkeys are generally found in high-density forests where they forage on leaves, flowers and fruit, but mostly leaves. Unlike other primates, their digestive system is similar to hoof animals with a special fermenting stomach. This enables the colobus monkey to exist in habitats where other primates cannot survive.

It is not unusual for a female to care for any child and they are non-aggressive towards each other. Angolan black and white colobus monkeys develop complex group and family behaviors including personal greetings and group sleeping patterns

The colobus monkey has black fur and a black face, surrounded by long, white locks of hair. It also has a mantle of white hair on the shoulders. The long, thin tail can be either black or white, but the tip is white. They are about two feet long and tall and weigh between 20 to 40 pounds.

They are from the more southern basins of the Congo River in the Democratic Republic of Congo and although named "Angola" that is one country that they are rarely found. They can also be found in forests and mountains in Tanzania, Uganda, and Kenya. The scientific name of the Angolan Black and White Colobus Monkey is Colobus angolensis.

Okapi

Although the Okapi might look like a zebra with those stripes, it is actually related to the giraffe. They are often called the forest giraffe. The males have similar

skin-covered horns like the giraffe. Both the male and females have a long, prehensile black tongue. Prehensile means that the tongue is able to grasp, the way a monkey can use its prehensile tail to swing between trees. This tongue is so long that they will use it to groom their eyes and ears.The Okapi likes to eat fruit, tree leaves, grasses, and ferns. It has large and flexible ears and a long neck. Both the giraffe and the okapi include a similar walk. Both have a "pacing gait," stepping simultaneously with the front and the hind leg on the same side of the body. The Okapi population continues to decline due to poaching and habitat loss from commercial logging and mining. The Okapi can be found in dense rainforests of the Democratic Republic of the Congo. They are so reclusive that they are rarely encountered in the wild.

When you are lucky enough to see these at Disney's Animal Kingdom, they are usually alone. But if you do see two of them together it is normally a mother and her calf. As an endangered species, Disney has been trying to breed these under the guidelines of the Species

Survival Plan. They were successful when Zelda was chosen to mate with Mandazi. A new calf was born. A baby calf can stand up within 30 minutes of being born.

You can find the Okapi on the Gorilla Falls Trail, on the Kilimanjaro Safari ride, and at the Animal Kingdom Lodge.The scientific name of the Okapi is Okapia johnstoni.

Yellow-backed Duiker

Sharing the exhibit with the Okapi is a newer antelope addition to the Animal Kingdom in 2022. The yellow-backed Duiker live in forests and are native to Central and Western Africa from Senegal to Western Uganda with some in Gambia. They can also be found in Rwanda, Burundi, Zaire, and most of Zambia.

They prefer to eat fruit, but will also eat shoots, buds, leaves, and plant roots. Unlike other antelopes that live in herds, these prefer to live alone or as a couple.

The scientific name of the Yellow-backed Duiker is Cephalophussilvicultor.

Further up the trail, you'll come to a small building that is themed as a Research Center to study the animals in the area. In their research office you'll get to see several different animals in terrariums like lizards, snakes, and frogs and some tiny naked mole rats.

Naked Mole-Rats

It might be hard to find this tiny animal in the dark room; the naked mole-rat lacks pain sensitivity in its skin, so it stays underground. Look for the glass exhibit across the room where you can see them in their burrows.They are native to the tropical grasslands of East Africa, predominantly southern Ethiopia, Kenya, and Somalia. They only weigh a little over an ounce and are about 3 inches long. Like ants, there is a single queen that can be close to 3 ounces.

They eat large tubers that they find in their burrows underground. One tuber can provide their colony with enough food to last for months. They leave the outside of the tuber enabling it to regenerate. They sometimes also eat their own feces. This may be a way they can share hormones from the queen. The scientific name of the Naked Mole-Rat is Heterocephalusglaber.

After the research center is an aviary with many fish and birds that I will discuss in future books. You might spot a hippo in a fish pool getting "cleaned." The full hippo description is in the safari ride section.

Meerkats

You'll find these adorable little meerkats on the Gorilla Falls Trail between the bird aviary and the gorilla family exhibit. Meerkats are found in the deserts and grasslands of Africa, including southwestern Botswana, western and southern Namibia, and north and west South Africa.

During one of our visits, all the meerkats were up at Rafiki's Watch getting their regular medical checkup. With such a strong social network, to make sure one of the meerkats isn't rejected from the group; they take all of them at once for their medical checkups. Weighing about 2 pounds, meerkats have tough foreclaws for digging and have the ability to regulate their internal body temperature to survive in their hot and dry habitats. The dominant female can be a little heavier than the others. The meerkat's ears can be closed to protect them during digging.

They are very social animals and from 2005 to 2008, the show Meerkat Manor gave us a glimpse into life in Africa's Kalahari Desert through the eyes of a family of meerkats. A new show was on again in 2021 on BBCAmerica. A typical group is between a dozen and thirty meerkats with the older and heavier ones being dominant. There is usually a sentry posted to alert the others to danger with a variety of vocalizations to tell the others specifics about the danger.

Meerkats live in rock crevices and in large burrow systems with often over a dozen entrances and up to three levels. They eat mostly insects, but also like fruits and vegetables. This group of meerkats are trained to go to their feeding station to make sure everyone gets a fair amount of food each day. They are the most active in the early morning and the cool afternoon. The scientific name for the Meerkat is Suricata suricatta.

Western Lowland Gorilla

Disney's Animal Kingdom features a family group of western lowland gorillas called a troop and some bachelor males. In episode 8 of Disney+'s "Magic of Disney's Animal Kingdom," from National Geographic, the baby,

Grace worked with therapists to get better use of her arm. In episode 2, Gino - the father of the family troop had a birthday cake made of fruit and sweet potatoes that he ended up sharing with his daughters and sons.

The western lowland gorilla eats shoots, fruit, wild celery, tree bark and pulp which is abundant in the thick forest of Central and West Africa. During the wet season gorillas commonly consume fruits. They may also eat insects from time to time. The males are often called silverbacks because the hair on the back and rump of males takes on a grey coloration and is also lost as they get older.You can find the gorillas on the Gorilla Falls trail. The family is normally at the first viewing area and then a few of the older males are in the other areas close to the bridge.The scientific name for the Western Lowland Gorilla is Gorilla gorillagorilla

Kilimanjaro Safari

The safari ride often starts out with the **Okapi**, which is also on the left of the Gorilla Falls Trail. You can see the information about the Okapi in the earlier Gorilla Falls section. The rest of the animals are roughly in the order they appear; but often they are hidden during your ride – or moved to another area.

Black Rhinoceros

The black rhinoceros is native to eastern and southern Africa including Angola, Botswana, Kenya, Malawi, Mozambique, Namibia, South Africa, Eswatini, Tanzania, Zambia, and Zimbabwe. You can see this

majestic animal on the safari ride. They are normally before the white rhinos and to tell them apart, the black rhino has a hooked lip while the white rhino has a square lip. White rhinos have a longer skull and a bigger shoulder hump.

Although the rhinoceros is referred to as black, its color can vary from brown to grey. The distinctive rhino horn is made up primarily of keratin – a protein found in hair.They are considered a "browser" and will eat fruit, leafy plants, branches, shoots, and even thorny wood bushes. In episode 5 of Disney+'s "Magic of Disney's Animal Kingdom," from National Geographic, Badru gets some blood drawn. The scientific name of the Black Rhinoceros is Dicerosbicornis.

Bongo Antelope

The Bongo are considered shy, but it is an easy animal to spot with their beautiful reddish-brown coat with black and white markings. Both the male and females have long slightly spiraled horns. Between the eyes there is a white marking and two large white spots grace each cheek. Another white marking occurs where the neck meets the chest. Bongos have no special secretion glands, so these markings might help them identify each other. The lips of a bongo are white, topped with a black muzzle. They have a long prehensile tongue which it uses to grasp grasses and leaves.

Bongos like to feed on leaves, bushes, herbs, vines, bark, grasses and fruits. They need salt in their diets, and are known to eat charcoal left from trees that have burnt. Their large ears help them hear predators.

Native to Africa, they are the third-largest antelope in the world. They are found in tropical jungles with dense undergrowth in Central Africa, with some populations in Kenya. They are also found in Cameroon, the Central African Republic, the Republic of the Congo, the Democratic Republic of Congo, the Ivory Coast, Equatorial Guinea, Gabon, Ghana, Guinea, Liberia, Sierra Leone, and South Sudan. In 2005, three of Animal Kingdom's bongos were relocated to Kenya as part of an international effort to return the nearly extinct creature to its native land. The scientific name for the Bongo is Tragelaphus eurycerus.

Greater Kudu

The greater kudu is one of the largest antelopes and is found in eastern and southern Africa. You can see them in the savanna of the safari ride. The males have large horns with 2 ½ twists that would be almost four feet if straightened. Both the male and female have a narrow body and long legs. Their coats vary in degrees of bluish grey to red/brown. They have up to 12 vertical white stripes along their torso. The head can be darker than the rest of the body, and has a white chevron between the eyes. Their ears are particularly large and broad for the size of their head and fringed with white hair.

The greater kudu are grazers and love to feast on fruits. They are considered to be browsers that eat leaves and shoots from a variety of plants. In dry seasons, they eat wild watermelons and other fruit for the liquid they provide. The lesser kudu is less dependent on water sources than the greater kudu.The scientific name of the Greater Kudu is Tragelaphus strepsiceros.

Hippopotamus

The hippopotamus is native to sub-Saharan Africa. They are the 3rd largest land animal after the elephant and rhinoceros. Back in February 2018, a baby boy hippo was born at Disney's Animal Kingdom. Disney Cast Members named the sweet boy Augustus or "Gus" for short. In episode 8 of the Disney+'s "Magic of Disney's Animal Kingdom," from National Geographic, Gus made a 'frenemy' out of an older hippo at the river playground to enable him to socialize with other hippos. Later we learned that Gus has joined a new family at the Dallas Zoo to help with the Association of Zoos and Aquariums' Species Survival Plan.

You can see the hippos both on the Gorilla Falls Trail in the fish tank where tiny fish "groom" the hippos, or on the safari ride in the large pond. To stay cool in the heat, hippos spend most of their day in rivers and lakes. Their eyes, nose and ears are located on the top of their head, which means they can see and breathe whilst submerged in the water. Hippos sweat an oily red liquid which helps protect their skin from drying out and acts as a sunblock.

Hippos eat around 75 lbs. of grass each night and will travel about 5 miles in a night to graze. They also eat fruit that they find along their trip. They can weigh from 3,000 to over 9,000 pounds.The scientific name for the Hippopotamus is Hippopotamus amphibius.

Nile Crocodile

During episode 5 of the Disney+'s "Magic of Disney's Animal Kingdom," from National Geographic, Mr. Campbell, a massive Nile crocodile, came in for a checkup for his injured tail, a result of a dust-up with another crocodile. The crocodiles are trained to come with the tap of a metal bar hit underwater. These massive animals can be seen on the safari ride, as you cross the bridge. They are native to freshwater habitats in at least 25 countries of Africa. They are capable of living in saltwater, but are only occasionally found in salty deltas and brackish lakes. They can be up to 20 feet long and weigh over 900 pounds. With their thick, scaly, and armored skin, they are considered apex predators. Crocodiles are

considered very aggressive but can wait for hours, days, and even weeks for the right moment to attack. They will eat almost any type of prey including fish, reptiles, birds, and mammals. The scientific name for the Nile Crocodile is Crocodylusniloticus.

Ankole Cattle

This giant-horned cattle are native to Uganda, Rwanda, and Burundi and the Nkole tribe's Sanga variety in Uganda is known as the Ankole. In Rwanda and Burundi, the Tutsi tribe's Sanga variety is called the Watusi. The Rwanda common strain of Watusi is called Inkuku.

Four of the cattle; Ace, Audrey, Dixie-Jane and Adeola were featured in background shots of Disney+'s "Magic of Disney's Animal Kingdom," from National Geographic. They can be seen at Disney's Animal Kingdom in the savanna on the safari ride.Traditionally, Ankole-Watusi were considered sacred. They supplied milk to the owners, but were rarely used for meat production, since an owner's wealth was counted in live animals. Cows weigh between 900 - 1200 pounds and bulls can weigh as much as 1600 pounds. The horns can be up to 8 feet from tip to tip but are not too heavy because they are hollow. They are normally solid in color, but can be spotted too. Rust is a common color. Like domestic cattle, they are grazers and feed on grasses.The scientific name for Ankole Cattle is Bos taurus indicus.

Giraffe

The giraffe is the tallest living terrestrial animal at 16-19 feet. Some think the T-Rex was taller, but others think that the museums mount the dinosaurs to make them appear taller. Regardless, the giraffes are tall. This long neck enables them to eat leaves, fruits and flowers of woody plants, primarily acacia species, which other animals cannot reach. Their hearts can weigh 25 pounds and are 3 inches thick to pump all the blood up to their brain.

Disney's Animal Kingdom has two subspecies of giraffes. I saw the Masai giraffes on the savanna during the safari ride, while the Reticulated giraffe, Hybrid giraffe and Masai giraffe are at Disney's Animal Kingdom Lodge. Their twenty-inch purple prehensile (capable of grasping) tongue enables them to feed on a range of different plants and shoots. It is thick and that protects it from thorns. In episode 1 of Disney+'s "Magic of Disney's Animal Kingdom," from National Geographic, Kenya, a Masai giraffe gets her hooves trimmed, by Dr. Dan.

Giraffes can sleep standing up. They bend their neck backwards and rest their head on the hip or thigh, a position believed to invoke short "deep sleep." They have an unusual walk. They move the legs on one side of the body, then do the same on the other side. When they run their hind legs move around the front legs before the latter move forward and the tail curls up.The scientific name for the Giraffeis Giraffa camelopardalis.

Wildebeest

The wildebeest is an antelope native to Eastern and Southern Africa. I have seen them both at the Animal Kingdom Lodge and on the savanna of the safari ride. They are easy to spot with their large horns and muzzles. They have a shaggy mane and tail. The ones I saw were blue wildebeest that have a blue/grey doat and weigh about 500 pounds. Both the male and female have those curved horns. They like to graze on grass alongside other herds like zebras to afford them protection against lions, wild dogs and other predators. Compared to other grazers that like dry, longer grasses, the wildebeest prefers the sweet and stocky grasses. These grasses often grow where there have been recent fires. The scientific name of the Blue Wildebeest is Connochaetestaurinus.

African Wild Dog - Painted Dog

The African Wild Dog is native to Africa and is the largest wild canine in Africa. You can see them normally in the mornings in the same exhibit as the hyenas occupy in the afternoon on the safari ride.Riddler is one of the African Painted Dogs that was featured on an episode of Disney Animals on Disney Junior in 2018 during their species highlight.They have big ears and unusual "calico" types of coats of black, white and golden brown. The puppies are born black and white and develop the golden part as they grow up. These patterns are unique to each dog and help identify them from distances far away. They weigh about 50 pounds, but can get as heavy as 70 pounds. Females are slightly smaller and they will leave the pack instead of the males that normally stay with the same pack.

The African Wild Dogs live in packs and are very social. They hunt during the day and eat mostly antelopes. They will also hunt other animals like warthogs and the calves of larger animals like African buffalo. They can run around 40 miles per hour while chasing prey. The scientific name of the African Wild Dogis Lycaon pictus

Spotted Hyena

In 2015 two hyenas, Scooter and Zawadi were added to the safari ride to share the exhibit with the painted dogs. Normally the hyenas are in the afternoon. They might look like dogs, but are closer to cats genetically. In episode 4 of Disney+'s "Magic of Disney's Animal Kingdom," from National Geographic, the pair got a bath. Hyenas live throughout much of Africa and eastwards through Arabia to India, but the spotted hyena are native to sub-Saharan Africa.

Hyenas are skilled hunters that can take down wildebeest or antelope, but also eat fruit, birds, lizards, snakes, and insects. They have powerful jaws that have 40% more force than a leopard.

Their rear is rounded which prevents attackers coming from behind from getting a grip on it. The hyena's head is wide and flat with rounded ears rather than the pointed of striped hyena. Each foot has four digits, which are webbed and armed with short, stout and blunt claws; like a cat. They have a short tail about a foot long. The female spotted hyena is considerably larger than the male. The scientific name for the Spotted Hyenais Crocuta crocuta.

Sable Antelope

The sable antelope is one of the most common antelopes on the safari ride. The population is doing quite well after five calves were born in 2015 alone with three generations growing and doing well.

The sable antelope are native to East and Southern Africa, all the way from Kenya to South Africa. There are also some in Angola.

The males are larger than the females at over 500 pounds while the females are in the 400-hundred-pound range. Both sexes have ringed horns which arch backwards. The horns are over 3 feet long in the females and the male horn can get to be over 5 feet long. The male can have a thick mane along the back and throat. This mane is a rich chestnut brown that turns darker; almost black for older males. They have white markings on their face, belly, and bottoms.

Like many other hoof stock, the sable antelope feeds upon foliage, grasses, leaves and herbs. They are most active in the early daylight, and less active during the hottest part of the day. The scientific name of the Sable Antelope is Hippotragus niger

Zebra

In episode 3 of Disney+'s "Magic of Disney's Animal Kingdom," from National Geographic they introduced zebras into the Kilimanjaro safari ride on the savanna. Two of the Hartmann's Mountain zebras, Clementine, and mom-to-be Prima were very interested in the process. Zebras can also be seen on the Gorilla Falls Trail behind the meerkats. Disney has three varieties; Grevy's zebras, Hartmann's Mountain zebras and plains zebras. Grevy's zebras are on the Gorilla Falls Exploration Trail. Both Hartmann's Mountain and plains zebras are at Disney's Animal Kingdom Lodge.

Although they look like a horse, zebras are one of the most recognized animals by their bold black-and-white striping patterns. The belly and legs are white when unstriped, but the muzzle is dark and the skin underneath the coat is uniformly black. You can tell the Grevy's and Hartmann's Mountain zebras apart from the plains by the lack of stripes on its belly. Zebras like to eat grasses and can survive on lower-quality vegetation.The scientific name of the Plains Zebra is Equus quagga. The Hartmann's Mountain Zebra is Equus zebra hartmannae. The Grevy's Zebra is Equus grevyi.

Eland

The Eland has the distinction of being the slowest antelope, reaching only 25 miles per hour. But if they take their speed down towards 15 mph, they have great endurance. The eland can jump; an amazing 8 feet from standing still. Their horns are twisted and have ridges. The males are furrier with a mane and then fur on their foreheads and also long the folds below their neck.

During Hurricane Dorian in 2019 a cute male was born and named Doppler after the radar system that provides great tracking of dangerous storms.

The Eland is native to Angola, Botswana, the Democratic Republic of the Congo, Eswatini, Ethiopia, Kenya, Lesotho, Malawi, Mozambique, Namibia, Rwanda, South Africa, South Sudan, Tanzania, Uganda, Zambia and Zimbabwe. The scientific name of the Eland is Taurotragus oryx.

Mandrill

The mandrill is native to western Africa, and can be found mostly in southern Cameroon, Gabon, Equatorial Guinea, and Congo. In episode 4 of Disney+'s "Magic of Disney's Animal Kingdom," from National Geographic, they added a male, Linus, into the all-female mix of mandrills that you can see on the safari ride. Although they look like baboons, they are actually part of the old-world monkey family. In April 2021, a new baby girl was born, named Ivy.

The males weigh around 80 pounds with the females about half of that. They have green or grey fur with yellow and black bands and a white belly. Their hairless face has a long muzzle with red stripes down the middle and blue ridges on the sides. The mandrill also has red noses and lips, a yellow beard and white tufts. Their bottoms can be many colors; red, pink, blue, scarlet, and purple. All the colors of the mandrills are more pronounced in dominant adult males.

They like to eat mostly fruits and insects. Mandrills will also eat eggs, and even small vertebrates such as birds, tortoises and frogs. They sleep in trees at a different site each night. The scientific name of the Mandrillis Mandrillus sphinx.

African Elephants

These gentle giants are seen twice during the safari ride; the males are normally separate from the female dominated family herd. During episode 7 of Disney+'s "Magic of Disney's Animal Kingdom," from National Geographic, one teenage, Nadirah has a fear of crossing a bridge that connects two yards of the elephant habitat. Baby sister, Stella is also featured.

The African bush elephant is the largest land animal. Male elephants can weigh a little over 7,000 pounds and be 13 feet tall. The African bush elephant's back is concave-shaped, while the back of the African forest elephant is straighter. Compared to the Asian elephant, the African elephant's ears are larger. An elephant's ears are made up of a complex network of blood vessels, which help with regulating an elephant's temperature. Blood is circulated through their ears to cool them down in hot climates. They also use their ears to communicate visually. Flapping their ears can signify either aggression or joy.

The African elephant eats various plants and unusual parts of the plant, including the stems, bark and roots. They can eat close to 200 pounds a day and drink 50 gallons of water. The scientific name of the Elephantis Loxodonta africana.

Springbok

In 2013, Disney added six female springboks to the savanna on the safari ride at Disney's Animal Kingdom. These medium-sized antelopes normally weigh between 50 and 90 pounds and are native to southern and southwestern Africa.

They are brown/rust colored with a white belly and face and dark area running along the top of the belly. Both the male and female have horns that curve backwards. They can jump straight up, around 6 feet with all four of their legs stiff. The South African rugby team gets its name from this cute antelope.

The springbok likes to eat from bushes and can go years without water, getting their fluid from succulent plants. They are the most active during dawn and dusk. The scientific name of the Springbokis Antidorcas marsupialis.

Lion

Lions were some of the stars of new Disneynature film, "African Cats," which premiered in 2011, especially the female cub, Mara. The lion is native sub-Saharan Africa and western India, but you can see them usually sleeping high on top of the large rocks from the safari ride. These rocks have air conditioning and heating vents, so it is a comfortable place for a nap.

I have sketched the male lion who is larger than the females and has a big mane. The females hunt together and can weigh up towards 300 pounds compared to 500 pounds for a male. They can run quickly, but only for short distances.

They are considered an apex predator, feeding mostly on hoof stock on the savannas like zebra, giraffe, and gemsbok. At Disney's Animal Kingdom on a hot summer day instead of their regular meal of raw meat, they will get some frozen fish or meat in a block of ice. To simulate stalking behaviors and provide exercise, big plastic balls are stuffed with the lions' favorite scents; cinnamon and pumpkin pie. The scientific name of the Lion is Panthera leo.

Cheetah

A cheetah named Sita was one of the stars of the Disneynature film, "African Cats," which premiered in 2011. On the safari ride, right after the savanna, you can see some of the current cheetahs at Disney's Animal Kingdom.

These cats are native to Africa and central Iran and live mostly on savannas. The cheetah is the fastest land animal and can reach speeds of up to 80 mph. They can weigh as much as 150 pounds or more and are about 3 feet tall. They have a tan coat evenly spaced black spots.

They eat mostly hoofed animals like springbok, gazelles, and impalas. They have been known to eat fruit like melons for the water. Cheetahs hunt during the day, often at the cooler times at dawn and dusk. Big eaters, they can eat about 20 pounds in two hours. The scientific name for the Cheetahis Acinonyx jubatus.

Southern White Rhinoceros

In episode 6 of Disney+'s "Magic of Disney's Animal Kingdom," from National Geographic, Dugan had a birthday party. But the star lately has been his soon, Ranger. His mom, Kendi was paired with dad Dugan through one of the Species Survival Plans overseen by the Association of Zoos and Aquariums to ensure the responsible breeding of endangered species.

Most southern white rhinoceros are in the south African countries; South Africa, Namibia, Zimbabwe, Kenya and Uganda. They are one of the largest land animals, weighing between 3,000 and 5,000 pounds. They have two horns; the front one can be two feet in length. They get their name "white" which is close to the word,"wide" for their wide mouths used for grazing. They drink twice a day if water is available, but can live up to five days without water. The southern white rhino spends about half of their day eating and one third resting and like sitting in mud holes to cool down. Their scientific name of the Southern White Rhinoceros is Ceratotheriumsimumsimum.

Blesbok and Bontebok

Out on the Kilimanjaro Safari Savanna, there are many animals that are considered "hoof stock" which are mammals that chew the cud regurgitated from its rumen. These include cattle, antelopes, giraffes, and the blesbok. The blesbok can be found in the wild in both South Africa and Namibia.

The blesbok has a reddish upper body and ridged horns. Their preferred habitat is open grassland with water. They were once many of them on the African plains, but have become scarce since the late 1800s due to relentless hunting for their skins and meat. They are very shy and alert; and fast. They can maintain speeds of over 40 miles per hour. They don't jump well, but can duck under things. I have a lot of trouble telling all the antelopes on the savanna apart; but the blesbok has a large white patch with a horizontal brown stripe which divides this patch above the eyes. In addition, the horns have rings. The neck and the top of the back of the blesbok is brown but as you look down their coat, the coloring becomes darker. There is more white patches on the belly up towards the tail. Look for ringed horns and that white patch on the face. The Afrikaans word for a blaze such as one might see on the forehead of a horse is "bles" - thus their name, blesbok.The scientific name of the Belesbok is Damaliscuspygargusphillipsi.

The bontebok is an antelope very similar to the blesbok that is now extinct in its natural habitats, but farming has brought back their numbers in South Africa, Lesotho and Namibia.

They are tall compared to other medium-sized antelope, standing over 3 feet high at the shoulder and can be 4 to 7 feet long. They are dark brown with a white belly, rump, tail and a white stripe on their forehead. They rest during the heat of the day and eat mostly grasses. A typical herd would be about 30-40.

Both the male and female bontebok have horns about 3 feet long that are ringed and have a twisting shape. According to Joe Christman, mammal curator at Disney's Animal Kingdom, "we had planned all along to display bontebok on the savannas, but have been unable to obtain them," Joe said. "The blesbok is a very similar species -- some experts say it's the same species -- and allows us to test the waters before obtaining true bontebok." They often interbreed between the blesbok and bontebok. The scientific name for the Bontebokis Damaliscuspygargus - very similar to the Blesbok.

Impala

The impala is an antelope found in eastern and southern Africa. You can find it at Disney's Animal Kingdom Lodge and sometimes on the savanna of the safari ride.

The males are easy to spot with their highly ridged and arch-like horns. The females don't have horns and weigh about 50 pounds less than the males. Both are considered to be medium sized of the antelopes. They are reddish brown with a white underbelly. They have a bushy white tail with some dark stripes on the tail and hindlegs. They can leap almost 10 feet.

Impala eat grasses as well as leaves and bushes - depending on the season and availability of each. They like to stay close to the water sources, and will eat succulent vegetation if water is scarce. The scientific name of the Impala is Aepyceros melampus.

Gemsbok

The gemsbok is an antelope native to the arid regions of Southern Africa, such as the Kalahari Desert. You can spot it most easily at Disney's Animal Kingdom Lodge and occasionally on the savanna of the safari ride.

The gemsbok is tan with a long black tail. They are big; the males can be 4 feet at the shoulder and weigh around 500 pounds. But it is their horns that really make them stand out. The male horns can be almost 3 feet in length. The female's horns can be longer, but are thinner.

The gemsbok mostly eats grasses but during the dry season, they can switch to the leaves on brushes when grass is sparse. It can also dig up to a couple feet deep to find roots and tubers. They get some of their water by eating melons and cucumbers, which can provide most of the water required. The scientific name of the Gemsbok is Oryx gazella.

Nyala

The nyala is an antelope found in southern Africa; Botswana, Malawi, Mozambique, Namibia, South Africa, Swaziland, Zambia, and Zimbabwe.I first heard about the nyala when the Mara restaurant at Disney's Animal Kingdom lodge introduced a Nyala Brownie.

The nyala can be spotted both in the savanna of the safari ride and at the Animal Kingdom Lodge. But if you see a female, it will look a lot different than a male. The

female that I have drawn doesn't have horns and its coat is a rusty orange color with vertical white stripes. These stripes are missing or faint in the male that has a greyer or tan coat. The male has large twisting horns that have a yellow tip. Both the male and female have a white spot between their eyes. The males have longer fur on their back and belly.

The nyala likes to eat leaves, fruits, flowers and twigs. During the rainy season they eat the fresh soft grass. They choose places to graze with a water source nearby since they need water. But they can survive in areas with only a seasonal availability of water. The scientific name of Nyala is Tragelaphus angasii.

Steenbok

In episode 6 of the Disney+'s "Magic of Disney's Animal Kingdom," from National Geographic, we saw Stark, a new baby steenbok, at Disney's Animal Kingdom Lodge just hours after he was born. But it was soon discovered that he had a problem with his legs. Like many of the hoof stock, the steenbok is seen both at the Disney's Animal Kingdom Lodge and the savanna of the safari ride.

The steenbok is a smaller antelope from eastern and southern Africa, standing less than 2 feet and weighing about 25 pounds. They have unusually large ears and the males have short horns about 6-7 inches long. They are a light orange/tan with while on their chin and underbelly.

Due to their short nature, they like to eat lower lying vegetation along with roots and tubers. They are known to also eat fruits and rarely graze on grass like other antelopes. They get most of the moisture they need from their food and don't need water.The scientific name of the Steenbok is Raphicerus campestris.

Thomson's Gazelle

Although this graceful animal is most commonly found at Disney's Animal Kingdom Lodge, it sometimes shows up in the savanna along with other hoofstock animals. It is native to East Africa, especially Tanzania and Kenya. and considered the common type of gazelle in East Africa.

They are a smaller gazelle, weighing about 50 pounds. You can spot them by that dark stripe going across their belly and the white circles around their eyes and two other dark patches on their forehead running from their eye to their nose. They have ringed horns that can curve as they get older.

The Thomson's Gazelle is fast; it can run upwards toward 55 mph and the only faster animals are the springbok, pronghorn and cheetah - which is a predator of this gazelle.They prefer shorter fresh grass, but during the dry seasons, they eat a variety of leaves from woody bushes. The scientific name of Thomson's Gazelle is Eudorcasthomsonii.

Warthog

I think the Disney film, *The Lion King* made the ugly warthog - loveable. Pumbaa, the warthog, was the best friend of Timon (a meerkat.) The warthogs at Disney's Animal Kingdom are one of the last animals on your left

as your safari starts to end. Native to sub-Saharan Africa, the warthog is found in grassland, savanna, and woodland. There are other similar pigs over at the Animal Kingdom Lodge, but they are the closely related rig river hogs.

They like to eat a variety of things depending on the season. They eat berries, grasses and roots along with bark, fungi, and insects.

Warthogs are powerful diggers and have two pairs of tusks that curve upwards. The lower pair can be razor-sharp from the warthogs rubbing against the upper pair every time the mouth is opened and closed. They are 2-4 feet tall and can weigh as much as 300 pounds. Warthogs can run speeds of up 30 mph when threatened and will run with their tails sticking up. They often enter their dens rear first with tusks facing out for protection. The scientific name of the Warthogis Phacochoerus africanus.

Nigerian Dwarf Goats

These adorable goats can be seen at the Affection Section at Rafiki's Planet Watch or as the last thing on the Kilimanjaro Safari ride on your left. They have created a simulated warden's outpost where the goats climb over the jeep, roof, and table. Native to West Africa, they actually evolved from the stocky West African Dwarf, but are smaller. They weigh about 75 pounds and are about 2 feet tall. They can be horned or hornless and their coat is short and fine. They are commonly gold, chocolate and black, frequently with white markings - but can be almost any color.

Goats prefer to eat vines and grasses like deer and although they will chew on almost anything, they are actually picky eaters. They are curious animals and like to sample things like cardboard boxes and clothing. The scientific name of the Nigerian Dwarf Goatis Capra hircus.

You are almost done; only one more animal left. If you are tired and hungry after the ride; three more quick service restaurants are available. Go straight toward the entrance and Pizafari on your right past Creature Comforts has indoor and outdoor seating and a great place to rest and get out of the cold, rain or heat. On a nice day, grab a shady, open-air table at Flame Tree Barbeque on your left towards Dinoland. There is spectacular food and views of the lake. But if you're looking for a climate controlled quiet refuge, you can't beat Restaurantosaurus. You'll also get to see the American Crocodile while you wait for your Mobile Order. Make your way across Discovery Island and look for the big sign for Dinoland.

Dinoland

If you go to the right towards the dinosaur ride, you'll see the one animal in Dinoland, the American Crocodile, which is on your left.

American Crocodile

This crocodile might look "fake" because it rarely moves and only sometimes opens its mouth to cool down. This is one of the few "American" animals in Disney's Animal Kingdom and is found in the wild throughout the "Americans" in South Florida and the coasts of Mexico. They have been found as far south as Peru and Venezuela. They like coastal rivers and lakes that have some salinity like brackish lakes, mangrove swamps, lagoons, cays, and small islands. You can find the American Crocodile in an exhibit across from Restaurantosaurus in Dinoland.

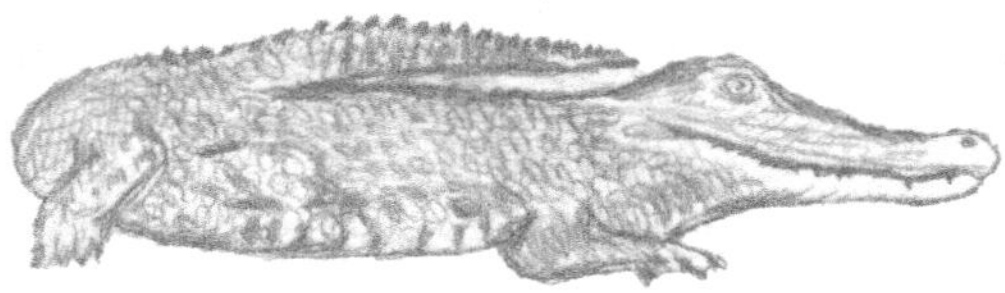

They are one of the biggest of the crocodiles; the males can get up to 20 feet and weigh 2,000 pounds. They have stocky legs, a long tail, and long snout with powerful jaws with a bite force of 3700 pounds. The nostrils, eyes, and ears are situated on the top of its head, so it can hide in the water to lure prey close.

The snout is longer and narrower than that of the American alligator. American crocodiles are also lighter and grayer than the darker American alligator. They can run after you at almost 10 mph on land and pursue you at 20 mph in the water. The American crocodile likes to eat small mammals, birds, frogs, turtles, and fish. The scientific name for the American Crocodile is Crocodylusacutus.

Animal Kingdom Lodge

If you're not ready for your day to end, but need a little rest or to get out of the cold, hot, or rain, exit the park to the right and get on the bus to the Animal Kingdom Lodge. You don't have to be a guest of the hotel to take the short ride and enjoy more animals. After the bus drops you off, enter the lobby and go straight to the rear and towards the huge picture windows, then down the stairs and out those doors below to the amazing views of the Arusha savanna. There are more viewing areas that I like, especially just past the hot tub. I often get something to eat at the quick service, The Mara and then eat it while watching the giraffes walk by. For a real treat, see if there are some Zebra cupcakes or Zebra Domes left for dessert. If you have time and do some planning, you can't miss with a buffet meal at Boma - Flavors of Africa - but you'll need to make reservations.

Ending your Walking Tour with Food

If you are meeting someone who has been riding the thrill rides all day, then ending this tour at Restaurantosaurus in Dinoland is a great meeting place. If you have lots more energy after you are finished with your safari ride, then you can take the long walk by turning right at the Tusker House Restaurant. The Tusker House is one of my favorite sit-down restaurants and has great food, but hard to get a reservation. Head towards The Lion King and deep into Pandora and meet them at the quick service, Satu'li Canteen. The food can be a little spicy, but the theming is amazing. Closer to the entrance on Discovery Island, The Flame Tree Barbeque and Pizzafari are more central and I would pick Pizzafari for the climate control indoors and Flame Tree for the food and ambiance; but it is outdoors and can be hot. My personal favorite is Yak and Yeti - the restaurant is amazing and the quick service cafe has a great breakfast and the restrooms are very close to your table. I normally start my day by walking the Oasis and Discovery Island and then have a breakfast bowl at Yak and Yeti quick service before tackling the two trails.

I hope this guide has helped you have an Amazing Day visiting with the animals at Disney's Animal Kingdom and look for my future guides where I'll share my adventures with the other animals, birds, and fish of

Disney World along with the plants, flowers and tree. If you are looking for a particular animal, the following alphabetical list should help you pinpoint its location.

Alphabetical List of Animals

Acknowledgements

Thank you to my editors at SmilingEagle Press, Cathy Phillips and Doogie Souz, for helping get this book completed with incentive and guidance.

A big thank you to Chandra who taught me how to write my first novel from one of my screenplays. Many thanks to my husband Keith who has given me love, encouragement, and constructive criticism for over fifty years.

A special thanks to my family and friends who critique my books and make comments; Laura Ailes, Gayle Higby, Amy Delk, Jennifer Ellis, Andrea Navarrette, Camille DeMoss, Jack Ross, Mark Saunders, JJ Waya, and Miss Michael.

About the Author

Sandi Jerome is a writer and graduate of UCLA's Advanced Screenwriting program. Her screenplay, ***Runaway Cricket***, is being produced as an animated musical by BlackOrb.com.

Sandi is an enrolled and blood member of the Cherokee Nation and a 2023 Native American Media Alliance fellow - twice. In her first NAMA TV writing fellowship, she wrote ***Technically Soccer***, a half-hour comedy about a Women's Professional Soccer team getting an AI-Robot coach. It is being developed into a TV series by Little Studio Films. Sandi is an avid women's soccer fan; she has coached, played, and refereed soccer for over twenty years.

Her middle-grade book, ***Sleep Warrior,*** about her Cherokee ancestor, is on Coverfly's Red List as the #3 Animated Manuscripts. Sandi's family is part of the Wolf Clan of the Cherokee tribe and she has written two screenplays with wolves. NOT werewolves - these wolves are real. ***Blood Moon Wolf*** (TV Pilot and Feature) was completed as part of her 2nd Native American fellowship about a wolf turning into the girl to be a spirit guide.

She grew up on an avocado farm in Escondido and was the "go to" kid to climb up high and pick the top fruit. She would then jump down into the thick pile of leaves and thought she could fly! She created a young adult fantasy book, ***Kira and Henry*** - where the teen princess has to hide the secret that she can fly or be banished from the kingdom. It was a 2nd Rounder in 2023 Austin Film Festival and a 2024 Kindle Book Review finalist.

Sandi is a scientist and loves designing edible gardens and doing botany experiments. Her early fascination with science and the nearby San Onofre Nuclear Generating station inspired her to write an action script, ***Use of Deadly Force***. Her husband was a technical writer in the nuclear industry and ensured its authenticity. Sandi aims to help emerging producers and directors get their projects to the screen.

She is a native Californian, born in Santa Ana. She belonged to a Science and Technology group and helped develop a volunteer Web on Wheels program that brought email to residents of assisted living centers. This inspired her body-switching comedy, ***Time for Lily***, which was a First-Round Finalist in Script Magazine Open Door Contest.

As a former computer programmer, she has been granted two patents for software design and recently sold her software company. She wrote a Sci-Fi, thriller script, ***Last Woman***, which involves the multi-universe and technology gone awry. It was a semifinalist in Final Draft's Big Break contest in 2023. Sandi partnered with Heidi Stangeland for the rewrite of this script and wrote another thriller with Heidi, ***Python Pursuit*** that now has a director attached.

Sandi was the first editor of Digital Dealer and wrote computer software reviews for major publications and numerous published computer guides.

As Floridian and a long-time Disney fan, annual passholder and certified Disney expert, Sandi wrote Pixie Dust Death set at Disney World, and then created a non-Disney version set at a theme park she invented. ***Wilma Wallaby Genius Girl Detective*** is being developed into a kid's series by a UK Production company. She is also the author of the ***Amazing Animals of Disney's Animal Kingdom***

Sandi wrote a book adaptation of **Hijacked** for producer Melissa Shevela of Helicopter Productions which is being shopped by fellow producer Autumn Bailey of AB Entertainment (*On a Wing and a Prayer* with Dennis Quaid.) Melissa has optioned Sandi's book, ***Cases of Nevada's Gaming Commission*** and they are developing a procedural TV series, ***Last Hand***. Her next book adaptation to film was ***Jake and Clara*** based on the book, Jake & Clara: Scandal, Politics, Hollywood and Murder by David R. Stokes. His previous book was optioned by Blair Underwood. They co-wrote a book very personal to Sandi, ***Churchill's Mum*** – about Jennie Jerome - the 4th cousin of her husband, Keith.

For Sandi, "***Writing is life***!"

Learn more at www.sandrajerome.com or leave comments at her publisher's Contact page, www.smilingeagle.com.

Made in the USA
Monee, IL
13 September 2026

62976604R00059